Nature Walk

Seeds

by Rebecca Stromstad Glaser

Bullfrog Books

Ideas for Parents and Teachers

Bullfrog Books let children practice reading informational text at the earliest reading levels. Repetition, familiar words, and photo labels support early readers.

Before Reading

- Discuss the cover photo. What does it tell them?
- Look at the picture glossary together. Read and discuss the words.

Read the Book

- "Walk" through the book and look at the photos. Let the child ask questions. Point out the photo labels.
- Read the book to the child, or have him or her read independently.

After Reading

- Prompt the child to think more. Ask: What seeds have you seen? How do seeds grow?

Bullfrog Books are published by Jump!
5357 Penn Avenue South
Minneapolis, MN 55419
www.jumplibrary.com

Library of Congress Cataloging-in-Publication Data

Glaser, Rebecca Stromstad.
Seeds / by Rebecca Stromstad Glaser.
p. cm. — (Bullfrog books: nature walk)
Summary: "Describing several examples of seeds, this photo-illustrated nature walk guide shows very young readers how to identify seeds and tells how the seeds spread. Includes photo glossary"
—Provided by publisher.
Includes bibliographical references and index.
ISBN 978-1-62031-029-8 (hardcover: alk. paper)
ISBN 978-1-62031-451-7 (paperback)
ISBN 978-1-62496-025-3 (ebook)
1. Seeds—Juvenile literature. I. Title.
QK661.G57 2013
575.6'8—dc23

2012009110

Series Designer: Ellen Huber
Book Designer: Ellen Huber
Photo Researcher: Heather Dreisbach

Photo Credits: All photos by Shutterstock except: Alamy, 9, 23tl; Dreamstime, 17, 20t; Getty Images, 10, 20b; iStockphoto, 5, 8b, 15, 16t, 19, 23br; Veer, 1.

Printed in the United States of America at Corporate Graphics in North Mankato, Minnesota.

Table of Contents

Looking for Seeds 4
Watch a Seed Grow 22
Picture Glossary 23
Index 24
To Learn More 24

Looking for Seeds

Let's go on a nature walk.

Do you see any seeds?

An acorn drops. Plop!

It will grow
an oak tree.

A maple seed spins.

It will grow
a maple tree.

burr

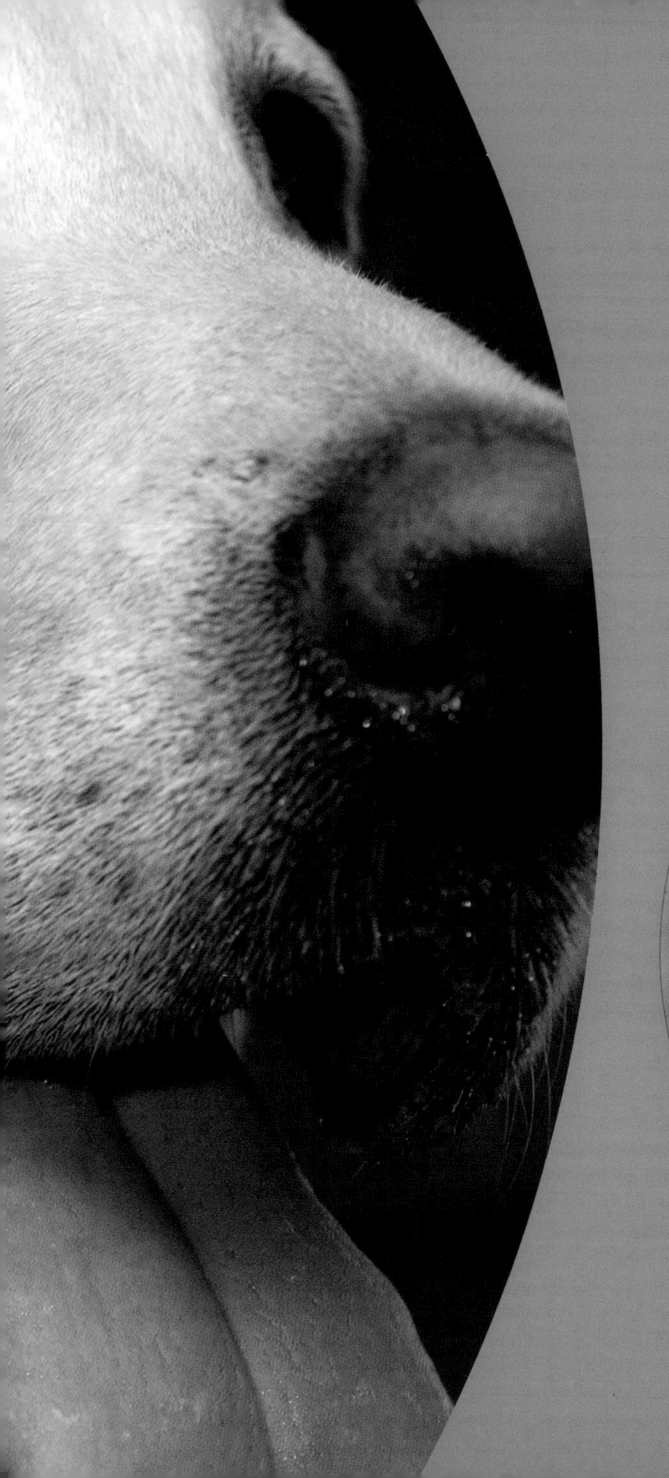

A burr sticks
to fur.

It will grow
a flower.

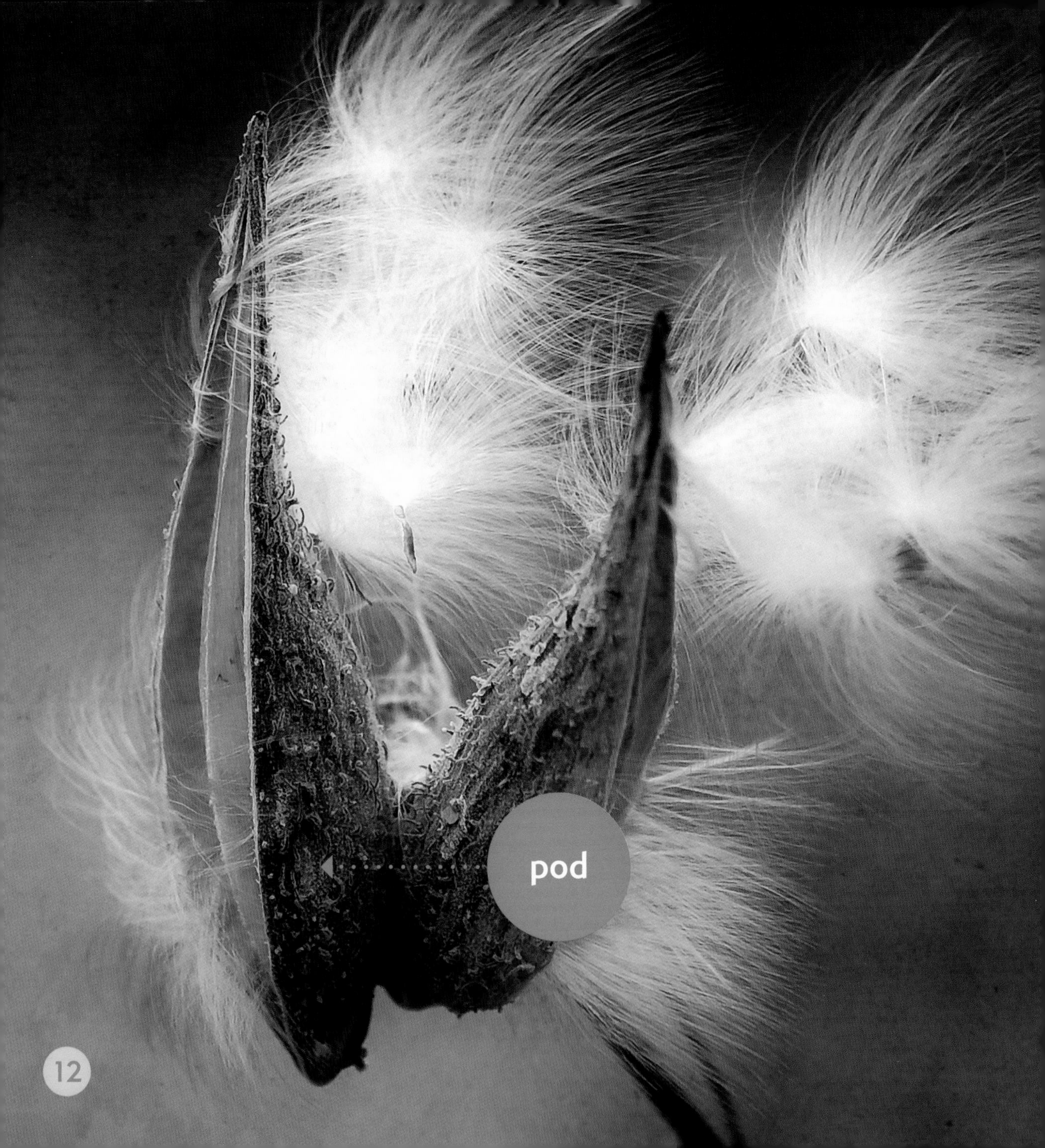
pod

A milkweed pod splits. Seeds blow away.

They will grow milkweed plants.

A dandelion seed blows in the wind.

It will grow a dandelion.

A lotus grows in water.
Its seeds float away.

They will grow a flower.

A pinecone opens.
Seeds are inside.

They will grow
a pine tree.

Not all the seeds will grow.
An animal may eat them first!